Nature's Mysteries

MINNESOTA'S DEVIL'S KETTLE

by Matt Lilley

abdobooks.com

Published by Pop!, a division of ABDO, PO Box 398166, Minneapolis, Minnesota 55439.

Printed in the United States of America, North Mankato, Minnesota.

102020
012021

THIS BOOK CONTAINS RECYCLED MATERIALS

Cover Photo: Shutterstock Images
Interior Photos: Shutterstock Images, 1, 12, 13, 19, 20, 23, 25; iStockphoto, 5, 6, 9, 16, 27; Don Douglas/Alamy, 8; Dominique Braud/Dembinsky Photo Associates/Alamy, 11; Tom Thulen/Alamy, 15; Red Line Editorial, 17; Harry Taylor/Dorling Kindersley/Science Source, 21; Tom Thulen/Alamy, 28

Editor: Alyssa Krekelberg
Series Designers: Candice Keimig, Victoria Bates, and Laura Graphenteen

Library of Congress Control Number: 2020940305

Publisher's Cataloging-in-Publication Data

Names: Lilley, Matt, author.

Title: Minnesota's Devil's Kettle / by Matt Lilley

Description: Minneapolis, Minnesota : POP!, 2021 | Series: Nature's Mysteries | Includes online resources and index

Identifiers: ISBN 9781532169199 (lib. bdg.) | ISBN 9781532169557 (ebook)

Subjects: LCSH: Judge C.R. Magney State Park (Minn.)--Juvenile literature. | Waterfalls--Minnesota--Juvenile literature. | Curiosities and wonders--Juvenile literature. | Mystery--Juvenile literature. | Geography--Juvenile literature.

Classification: DDC 910.02--dc23

WELCOME TO DiscoverRoo!

Pop open this book and you'll find QR codes loaded with information, so you can learn even more!

Scan this code* and others like it while you read, or visit the website below to make this book pop!

popbooksonline.com/devils-kettle

*Scanning QR codes requires a web-enabled smart device with a QR code reader app and a camera.

TABLE OF CONTENTS

CHAPTER 1

ONE RIVER, TWO FALLS

The Brule River rushes downstream. Ahead, there is a roaring sound. It's a waterfall. Water pours over the rocks. Half of the river falls 50 feet (15 m) and flows on. The other half falls into a hole and disappears.

WATCH A VIDEO HERE!

Devil's Kettle Falls is in Minnesota's Judge C. R. Magney State Park.

The Brule River is 40 miles (64 km) long.

The Brule River is in northern Minnesota. The river runs through a forest. It flows into Lake Superior. The waterfall is approximately 1 mile (1.6 km) from the lake. Rocks split the river in two. This pushes half the waterfall into the hole.

FUN FACT

The Ojibwa name for the Brule River is *Wiisaakode-ziibi*. This means "half-burnt-wood river."

Northern Minnesota has many hiking trails. One trail leads people right to Devil's Kettle Falls.

This strange waterfall is known as Devil's Kettle Falls. People can hike to see it. Scientists go there too. They have studied the waterfall. They wanted to know where the water goes after it falls down the hole.

WHERE IS DEVIL'S KETTLE FALLS?

CHAPTER 2

HOW THE FALLS FORMED

Geology helps scientists understand the mystery. Devil's Kettle was formed by heat and then ice. First came the heat. The waterfall's rocks are **volcanic**. They formed approximately

LEARN MORE HERE!

Some areas along Lake Superior have rhyolite cliffs.

1.1 billion years ago. The rocks are made of rhyolite. Rhyolite is hard. Water should not be able to tunnel through it.

Long after the rocks formed, **glaciers** moved through the area. Ice covered most of Minnesota. Glaciers scraped

Gravity forces glaciers to move.

Glacial movement helped create Lake Superior. Superior is one of the largest lakes in the world.

rocks and moved dirt. About 10,000 years ago, the glaciers melted. They left behind new hills, lakes, and rivers.

The rhyolite along the Brule River has cracks. Water got into the cracks and broke down the rock. The broken rocks got washed away. Eventually, a waterfall formed.

FUN FACT

Water can break down cracked rhyolite over time. However, water cannot make a large tunnel through rhyolite.

Rhyolite rocks are found all over Earth.

MIDCONTINENT RIFT

The rocks in Devil's Kettle formed along the Midcontinent Rift approximately 1.1 billion years ago. This rift was an 1,800-mile- (3,000 km) long crack in the earth. The rift almost split the continent apart to form an ocean. The Midcontinent Rift helped shape the area. Today, some old rocks left behind by the rift can be seen on Lake Superior's shoreline.

Basalt rocks are found along the Midcontinent Rift.

CANADA
N
W
E
S
LAKE SUPERIOR
NORTH DAKOTA
MINNESOTA
LAKE HURON
WISCONSIN
SOUTH DAKOTA
LAKE MICHIGAN
MICHIGAN
LAKE ERIE
IOWA
NEBRASKA
ILLINOIS
INDIANA
OHIO
KANSAS
MISSOURI
KENTUCKY
OKLAHOMA
ARKANSAS
Midcontinent Rift

CHAPTER 3

WHERE DOES IT GO?

For a long time, people didn't know where half the water from Devil's Kettle went. Some people thought the water went through a lava tube and came out in Lake Superior. Some **volcanic** rocks can form lava tubes. But rhyolite never forms them.

LEARN MORE HERE!

Lava tubes are underground tunnels carved by lava flow. They are sometimes called lava caves.

Many caves are made from limestone. Water dissolves this kind of rock easily.

Some people thought the water might be going through a cave. Sometimes, water can create caves.

Most caves form when water **dissolves** rocks. But water can't dissolve volcanic rocks. The rocks in the waterfall do have cracks, but the cracks are not big enough to carry so much water.

Rhyolite can be gray, blue gray, or shades of pink.

To find out where the water goes, some visitors tossed things into the river. They threw sticks or ping-pong balls. Then they would try to see where the objects came out. But nothing ever came back out.

FUN FACT

Waterfalls have strong currents. The current can hold down or destroy things dropped into it. That is why objects dropped into Devil's Kettle disappear.

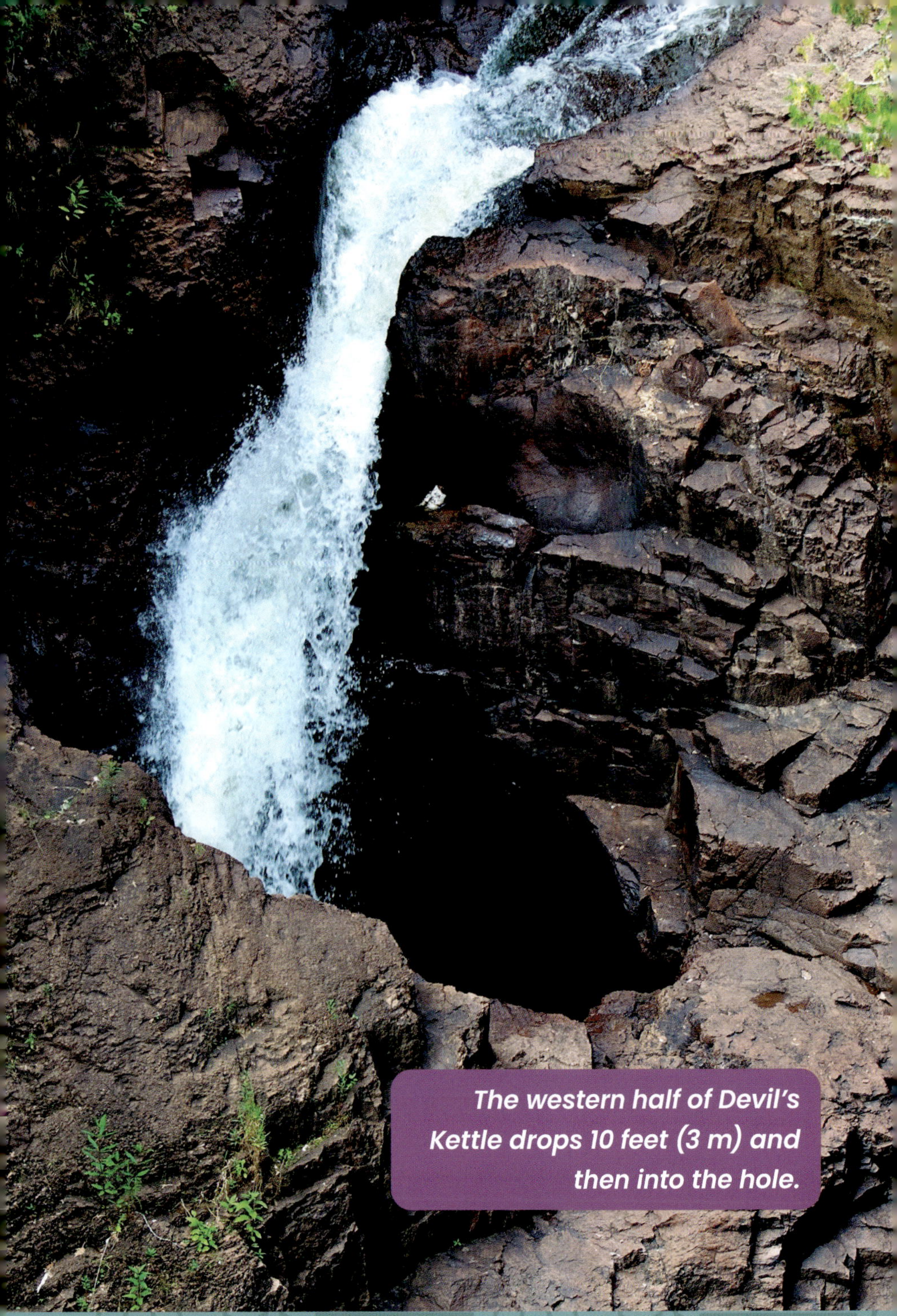

The western half of Devil's Kettle drops 10 feet (3 m) and then into the hole.

CHAPTER 4

SOLVING THE PUZZLE

Scientists kept searching for an explanation. Some of them wondered if the water was staying in the Brule River rather than going into Lake Superior. They decided to test this idea.

COMPLETE AN ACTIVITY HERE!

Scientists decided to study Devil's Kettle because they were curious about this natural feature.

Scientists did their study on Devil's Kettle in late 2016. They measured how much water was flowing in the river before it hit the falls. Then they measured how much was flowing after the falls. Measuring the water only took an hour or two.

MEASURING THE RIVER

Scientists had to measure several things on the Brule River in order to solve the mystery. First, they measured how wide the river was. Then they measured its depth in more than 25 different places. Finally, they measured how fast the water was moving in each place. With those measurements, they figured out the **volume** of the water flowing down the river.

Scientists who study water are called hydrologists.

Even though Devil's Kettle is no longer a mystery, many people still enjoy seeing the split waterfall.

The water's volume was the same in both places. That meant the water wasn't really going anywhere. It goes into the hole, but then it comes right back out into the river. The mystery of Devil's Kettle was finally solved.

FUN FACT

Someday scientists want to pour dye into the waterfall's hole. The dye would show them where the water comes back up in the river.

MAKING CONNECTIONS

TEXT-TO-SELF

Would you want to visit Devil's Kettle Falls?

Why or why not?

TEXT-TO-TEXT

Have you read any other books about scientists solving mysteries? What kind of scientists were they? What mysteries did they solve?

TEXT-TO-WORLD

Minnesota has been shaped by volcanic activity and glaciers. Can you think of other places that have had volcanic activity or glaciers?

GLOSSARY

current — part of a fluid body (such as air or water) moving continuously in a certain direction.

dissolve — to become part of a liquid such as water.

geology — a science that studies the history of the earth.

glacier — a large body of ice that moves slowly.

volcanic — made by a volcano.

volume — the amount of space that something takes up.

INDEX